Letts
gets you through

MATHS
QUICK PRACTICE TESTS

11+

Ages 10–11

11+
MATHS

FOR GL ASSESSMENT

QUICK PRACTICE TESTS

FAISAL NASIM

Contents

ACKNOWLEDGEMENTS

The author and publisher are grateful to the copyright holders for permission to use quoted materials and images.

Every effort has been made to trace copyright holders and obtain their permission for the use of copyright material. The author and publisher will gladly receive information enabling them to rectify any error or omission in subsequent editions. All facts are correct at time of going to press.

Published by Letts Educational

An imprint of HarperCollinsPublishers Limited

1 London Bridge Street
London SE1 9GF

ISBN: 9781844199150

First published 2018

10 9 8 7 6 5 4 3

© HarperCollinsPublishers Limited

British Library Cataloguing in Publication Data.

A CIP record of this book is available from the British Library.

Author and Series Editor: Faisal Nasim
Commissioning Editor: Michelle I'Anson
Editor and Project Manager: Sonia Dawkins
Cover Design: Paul Oates
Text and Page Design: Ian Wrigley
Layout and Artwork: Q2A Media
Production: Natalia Rebow
Printed in India by Multivista Global Pvt.Ltd.,

About this book

Familiarisation with 11+ test-style questions is a critical step in preparing your child for the 11+ selection tests. This book gives children lots of opportunities to test themselves in short, manageable bursts, helping to build confidence and improve the chance of test success.

It contains 22 tests designed to develop key numeracy skills. An example question and answer can be found at the start of Test 1.

- Each test is designed to be completed within a short amount of time. Frequent, short bursts of revision are found to be more productive than lengthier sessions.

- GL Assessment tests can be quite time-pressured so these practice tests will help your child become accustomed to this style of questioning.

- We recommend your child uses a pencil to complete the tests, so that they can rub out the answers and try again at a later date if necessary.

- Children will need a pencil and a rubber to complete the tests as well as some spare paper for rough working. They will also need to be able to see a clock/watch and should have a quiet place in which to do the tests.

- Your child should **not** use a calculator for any of these tests.

- Answers to every question are provided at the back of the book, with explanations given where appropriate.

- After completing the tests, children should revisit their weaker areas and attempt to improve their scores and timings.

Download a free progress chart, maths glossary and topic checklist from our website

letts-revision.co.uk/11+

Test 1

You have 10 minutes to complete this test.

You have 10 questions to complete within the given time.

Circle the letter next to the correct answer.

EXAMPLE

What is the area of this square?

5 cm

A	B	C	Ⓓ	E
20 cm²	10 cm²	30 cm²	25 cm²	50 cm²

---

① Which of these reads the same when rotated by 180°?

A	B	C	D	E
DHD	HAD	PNP	OHO	UBU

② Look at the function machine below.

$N \longrightarrow$ ×2 → −1 → ×4 → S

If N is input into the function machine, which of the following expressions represents S?

A	B	C	D	E
$8N - 1$	$2N - 4$	$8N - 4$	$2N - 1$	$2N + 1$

(3) Mark is trying to place 3 straws of different lengths to form a triangle.

Which of the following could be the lengths of 3 straws that Mark could use to make the triangle?

A	B	C	D	E
2 cm 2 cm 6 cm	4 cm 5 cm 10 cm	3 cm 4 cm 5 cm	3 cm 5 cm 9 cm	1 cm 2 cm 7 cm

(4) Stephen collects Japanese and Chinese coins. He has 440 coins in his collection, with 36 more Chinese than Japanese coins.

How many Chinese coins does he have?

A	B	C	D	E
232	238	403	212	180

(5) Shape A is made from 12 identical wooden cubes. The whole shape is painted blue.

How many of the wooden cubes have exactly 3 faces painted blue?

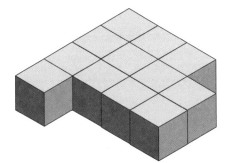

Shape A

A	B	C	D	E
2	3	4	0	5

(6) Neil buys 11 m of cord and cuts off 8 pieces, each with a length of n cm.
He has r cm of cord left.

Which of these expressions is equal to r?

A	B	C	D	E
$1100 - n$	$11 - 8n$	$1100 - 8n$	$3n$	$11 + 8n$

Questions continue on next page

(7)

Ryan recorded the number of stars he earned at school each week in the pie chart above. The pie chart shows how many times he earned each number of stars.

What was the mode of the number of stars he earned?

A	B	C	D	E
10	2	5	3	20

(8) The pictogram below shows the number of cakes ordered at a bakery over 5 days.

What is the mean number of cakes ordered per day?

Monday	
Tuesday	
Wednesday	
Thursday	
Friday	

Key:
◯ = 12 cakes

A	B	C	D	E
12	18	4	15	10

(9) If $9y - 7 = 29 - 3y$, what is the value of y?

A	B	C	D	E
3	2	4	0	8

(10) What is the perimeter of Shape F?

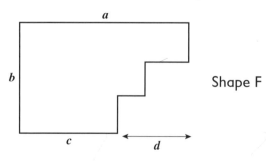

Shape F

A	B	C	D	E
$2a + b$	$2a + 2b$	$a + b + c + d$	$a + 2b$	$b + c + d$

Score: .... / 10

Test 2

You have 10 minutes to complete this test.

You have 10 questions to complete within the given time.

Circle the letter next to the correct answer.

(1) Which of the following has the greatest value?

A	30% of 120
B	$\frac{1}{2}$ of 80
C	Three-fifths of 90
D	52
E	0.2 of 250

(2) Laura books a hiking trip that costs £480.
She pays £20 weekly towards the cost of the trip. So far she has paid £60.

How many more weeks will it take her to pay the full cost of the trip?

A	B	C	D	E
24	22	21	20	23

(3)

8		4
	5	?
6		

This is a magic square in which the vertical columns, horizontal rows and two diagonals all add up to the same number.

Which number should replace the question mark?

A	B	C	D	E
1	2	7	9	3

Questions continue on next page

(4) Harry is now half of his sister's age. In 3 years' time, Harry will be 10.

How old will his sister be then?

A	B	C	D	E
20	17	14	13	7

(5) The spinner below is divided into equal sections. Sally spins the arrow on the spinner.

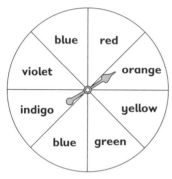

What is the probability that the arrow lands on a section that contains a word made of more than 4 letters?

A	B	C	D	E
$\frac{3}{8}$	$\frac{5}{8}$	$\frac{1}{2}$	$\frac{1}{4}$	$\frac{1}{3}$

(6) The sum of 3 consecutive numbers is 48. What is the largest number?

A	B	C	D	E
48	12	15	17	18

(7) A conversion chart for various ingredients is shown in the table below.

Cups	Weight in grams (g)
1 cup flour	150 g
1 cup sugar	225 g
1 cup butter	225 g
1 cup sultanas	200 g
1 cup almonds	110 g

A recipe uses 2 cups of butter, 3 cups of flour, one-fifth of a cup of sultanas and half a cup of almonds. What is the total weight of the measured ingredients?

A	B	C	D	E
900 g	995 g	1 kg	225 g	980 g

(8) Figure C is made from identical small squares that are shaded either black or white. What area of Figure C is shaded black?

Figure C

← 10 cm →

A	B	C	D	E
100 cm²	13 cm²	52 cm²	25 cm²	50 cm²

(9)

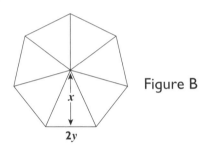

Figure B

x

$2y$

Figure B is a regular heptagon.

What is the total area of Figure B?

A	2xy square units
B	7xy square units
C	xy square units
D	8xy square units
E	14xy square units

(10) A map is drawn to a scale of 1:2000. What length on the map represents a real distance of 40 m?

A	B	C	D	E
2 cm	200 cm	20 cm	2 km	80 km

Score: .... / 10

Test 3

You have 10 minutes to complete this test.

You have 10 questions to complete within the given time.

Circle the letter next to the correct answer.

1 $(32 \times 56) - 40 = 1752$

Which of the following is correct?

A	$1752 + 40 = 32 \times 56$
B	$40 = 32 \times 56 + 1752$
C	$-40 = 32 \times 56 - 1752$
D	$32 \times 56 - 1752 = -40$
E	$32 \times 56 + 1752 = -40$

2 A soup recipe requires 500 g of tomatoes for every 200 ml of coconut milk.

Sheila only has 300 g of tomatoes. How much coconut milk should she use?

A	B	C	D	E
200 ml	120 ml	150 ml	50 ml	70 ml

3 Figure B shows the path of a remote-controlled toy car from start to finish.

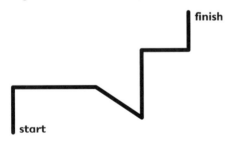

finish

start   Figure B

At how many points does the car turn through an obtuse angle?

A	B	C	D	E
1	2	3	4	0

④ Figure C is formed by arranging and cutting copies of the square and circle shown below. Each square has an area of a and each circle has an area of b.

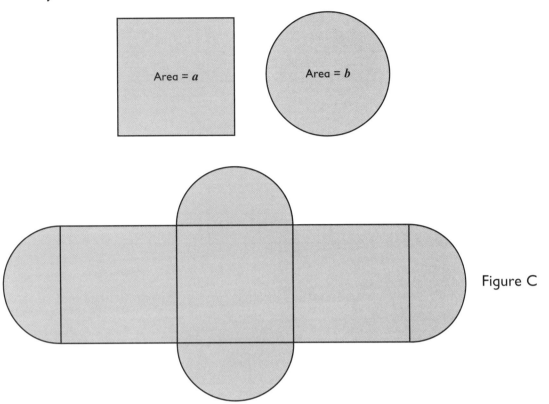

Area = a

Area = b

Figure C

Which expression shows the area of Figure C?

A	B	C	D	E
$3a + 4b$	$3a + b$	$3a + 2b$	$2a + 2b$	$2a + 3b$

⑤ The table below shows how many shirts of each pattern there are in a pack of 10.

Shirt pattern	Number
Dotted	3
Striped	4
Plain	2
Flowered	1

Emily picks a shirt at random. Which point on the scale below shows the probability that she will pick a striped shirt? Circle the correct letter.

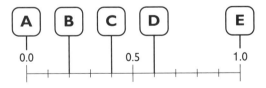

6   A rectangle has a perimeter of 28 cm. The width of the rectangle is 4 cm.

What is the area of the rectangle?

A	B	C	D	E
16 cm²	14 cm²	10 cm²	40 cm²	32 cm²

7   A number is first cubed and then divided by 1000. The final answer is 1.

What was the original number?

A	B	C	D	E
4	1	100	1000	10

8   Toby counted the number of packed lunches in each class at Bollow School and recorded his results in the table below.

Number of packed lunches	12	10	11	19	7
Number of classes	4	3	7	2	5

What is the mode of the number of packed lunches in a class?

A	B	C	D	E
19	11	7	12	10

9   A rucksack contains 400 matchboxes and weighs 11.8 kg. The empty rucksack weighs 600 g.

What is the weight of a single matchbox?

A	B	C	D	E
100 g	24 g	20 g	28 g	15 g

10 Jane writes a sequence in which the first two terms are 4 and 5.
She uses the rule below to find each of the next terms in the sequence:

Multiply the previous two terms and then subtract 5.

The first five terms are: 4   5   15   70   1045

What is the sixth term in the sequence?

A	B	C	D	E
10124	73150	75145	72145	73145

Score: .... / 10

Test 4

You have 10 minutes to complete this test.

You have 10 questions to complete within the given time.

Circle the letter next to the correct answer.

① One of the angles in an isosceles triangle measures 112°.

What is the value of one of the other angles in the triangle?

A	B	C	D	E
112°	68°	34°	180°	23°

② 40 children are going on a trekking trip. Each child brings x packets of food from home. Each child is also given y packets of food at school.

What is the total number of food packets carried by all the children?

A	B	C	D	E
$40x + y$	$40(x + y)$	$40x$	$40 + x + y$	$40xy$

③ The graph below shows a journey made by Andy.

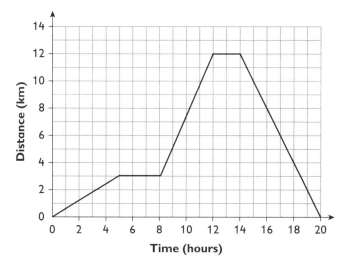

How much distance does Andy cover in total on this journey?

A	B	C	D	E
12 km	20 km	24 km	5 km	14 km

Questions continue on next page

4 A cotton bale weighs 331.1 kg. 119 kg of the cotton in the bale is sold.

What weight of cotton remains?

A	B	C	D	E
212.1 kg	212.9 kg	211.2 kg	339 kg	201.21 kg

5 Find the perimeter of the shape shown in Figure A.

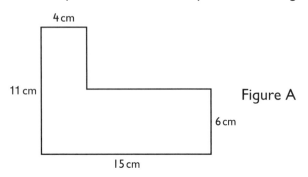

Figure A

A	B	C	D	E
50 cm	48 cm	52 cm	26 cm	38 cm

6 How many hours are there from 9 cm on Monday morning to 9 pm on Friday evening?

A	B	C	D	E
106	96	100	102	108

7 Figure B consists of 2 identical squares divided into triangles. The area of each of the shaded triangles in Figure B is s.

Which expression shows the unshaded area of Figure B?

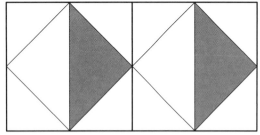

Figure B

A	B	C	D	E
$2s$	$4s$	s	$10s$	$6s$

(8) −11 < y < 5

Which of the following is a possible value of y?

A	B	C	D	E
10	−5	7	−12	15

(9) What is the order of rotational symmetry of this shape?

A	B	C	D	E
1	5	4	2	0

(10) The table below shows the relationship between x and y.

x	1	2	3	4	5	6
y	0	3	8	15	24	?

If x equals 6, what is the value of y?

A	B	C	D	E
17	29	41	35	10

Score: .... / 10

15

Test 5

You have 10 minutes to complete this test.

You have 10 questions to complete within the given time.

Circle the letter next to the correct answer.

(1) There are 52 cards in a pack. The cards are numbered from 1 to 52. One card is chosen at random.

What is the probability that the card chosen is odd?

A	B	C	D	E
$\frac{2}{13}$	$\frac{1}{13}$	$\frac{1}{2}$	$\frac{1}{52}$	$\frac{4}{13}$

(2) The regular pentagon below is rotated clockwise about its centre until point A aligns with the arrow at the top.

What angle does the pentagon turn through?

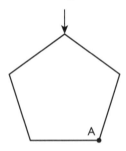

A	B	C	D	E
72°	144°	360°	200°	216°

(3) Which of the following is FALSE?

A	B	C	D	E
$\frac{3}{5} = 60\%$	$\frac{7}{20} = 35\%$	$\frac{3}{10} = 30\%$	$\frac{14}{50} = 28\%$	$\frac{4}{25} = 15\%$

(4) A cafeteria menu is shown below.

Minestrone soup	£3.50
Fruit salad	£2.75
Scrambled egg	£1.90
Coffee	£2.15

Yvonne buys two different items and pays with a £10 note. She receives £4.60 change.

Which two items did she buy?

A	Minestrone soup and Fruit salad
B	Minestrone soup and Coffee
C	Coffee and Scrambled egg
D	Minestrone soup and Scrambled egg
E	Coffee and Fruit salad

(5) Rich has 56 stamps. Three-sevenths of these are vintage stamps. A quarter of the remaining stamps are blue.

How many stamps are blue?

A	B	C	D	E
24	4	8	16	2

(6) Charles is facing south-east. He can see a house directly in front of him and a tree to his right.

In which direction must Charles face so that his back is facing the house?

A	B	C	D	E
south-west	north-east	south-east	north-west	south

(7) Paul thinks of a number, squares it and then divides by 4. The answer is 36.

What number did Paul think of?

A	B	C	D	E
9	144	12	11	28

Questions continue on next page

(8) The arrow below points to the weight of 4 bananas.

What is the weight of 6 bananas?

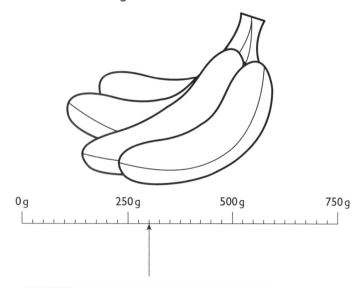

A	B	C	D	E
450 g	200 g	75 g	120 g	300 g

(9) A family of 5 go on holiday to Costa Rica. The total airfare is £2100, the total cost of accommodation and food is £700 and the total cost of day trips is £400.

What is the mean cost per person of these expenses?

A	B	C	D	E
£600	£3200	£800	£640	£520

(10) The star below has 10 internal angles.

How many of these internal angles are reflex?

A	B	C	D	E
10	0	4	3	5

Score: .... / 10

Test 6

You have 10 minutes to complete this test.

You have 10 questions to complete within the given time.

Circle the letter next to the correct answer.

① Which of these shapes has the greatest number of faces?

  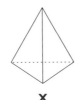

| V | W | X | Y | Z |

A	B	C	D	E
Z	W	X	V	Y

② The temperature inside a lab is −15°C and the temperature outside is 18°C.

What is the difference between the two temperatures?

A	B	C	D	E
33°C	3°C	18°C	20°C	25°C

③ A gardener charges £x per square metre to mow the lawn and a fixed charge of £10 to water and weed the plants.

What is the total cost of hiring the gardener to mow the lawn shown below and to water and weed the plants?

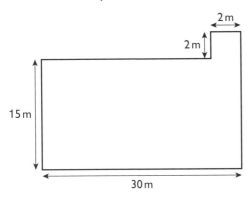

A	B	C	D	E
450x	£(4x + 10)	£(10 + 450x)	£(10 + 45x)	£(10 + 454x)

Questions continue on next page

(4) The four walls of a room, each measuring 4 m × 7 m, need to be painted. A tin of paint can cover a wall area of 12 m².

How many tins of paints are required to paint the four walls?

A	B	C	D	E
10	9	18	11	12

(5) Below is a regular hexagon.

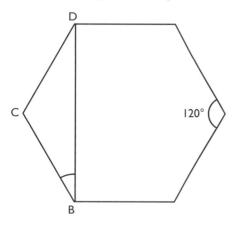

What is the value of Angle CBD?

A	B	C	D	E
120°	45°	30°	60°	25°

(6) Shelley recorded the time she spent walking in minutes on four different days in terms of n:

$2n + 1$ $3n - 4$ $n + 6$ $2n + 5$

What was the mean time Shelley spent walking per day in minutes?

A	B	C	D	E
$2n + 1$	$2n + 2$	$8n + 8$	$n + 1$	$2n - 1$

(7) Mrs Fletcher arranges 210 books in 4 piles. The first pile contains two-fifths of the books, the second pile contains a third of the books and the third pile contains 10% of the books. The fourth pile contains the remaining books.

How many of the 4 piles contain a number of books that is a multiple of 21?

A	B	C	D	E
1	2	3	4	0

(8) The dates of some important events are shown in the table below.

1972	Pocket calculators are invented.
1975	Microsoft is founded.
1976	Apple Computer Company is founded.
1977	The first Star Wars movie is released.
1981	Personal computers are invented by IBM.
1989	The World Wide Web is created.

Table A

Based on the information in Table A, which of these statements is FALSE?

A	The World Wide Web was created 12 years after the first Star Wars movie was released.
B	Microsoft and Apple Computer Company were founded in 2 consecutive years.
C	Pocket calculators were invented after personal computers.
D	The World Wide Web was created 1 year prior to 1990.
E	The first Star Wars movie is older than personal computers.

(9) $35728 \div 88 = 406$

What is $35728 \div 176$?

A	B	C	D	E
406	403	202	203	812

(10) A complete tank of water is drunk by 15 goats. The tank has a capacity of 96 litres.

How much water did each goat drink on average? Round your answer to the nearest litre.

A	B	C	D	E
6 l	6.4 l	7 l	9.6 l	96 l

Score: .... / 10

Test 7

You have 10 minutes to complete this test.

You have 10 questions to complete within the given time.

Circle the letter next to the correct answer.

1 How many of these shapes have an equal number of faces as they have vertices?

A	B	C	D	E
0	1	2	3	4

2 A cake recipe for 5 people requires 900 g of flour.

What quantity of flour (to the nearest kg) is needed to make enough cake for 14 people?

A	B	C	D	E
3 kg	2 kg	2.5 kg	4 kg	1 kg

3 Figure B below is formed from 3 identical hexagons and an equilateral triangle. 1 hexagon has an area of h cm^2.

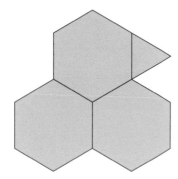

Figure B

What is the total area of Figure B in cm^2?

A	B	C	D	E
$3h$	$\frac{3h}{6}$	$3h + \frac{h}{6}$	$3h + \frac{h}{2}$	$18h$

④ What is the 11th term in this sequence?

5   11   17   23...

A	B	C	D	E
59	65	71	77	83

⑤ 900 people attended a marathon. 32% were adult females and 41% were adult males.

How many children attended the marathon?

A	B	C	D	E
288	369	243	321	323

⑥ The table below shows the amount of time William spent playing chess over 5 days.

Day	Start time	End time
Monday	9.15 am	9.45 am
Tuesday	3.20 pm	4 pm
Wednesday	1.15 pm	1.35 pm
Thursday	7.20 am	7.40 am
Friday	3 pm	?

If the mean time he spent playing chess per day was 25 minutes, what time did he finish playing chess on Friday?

A	B	C	D	E
3.10 pm	3.30 pm	3.15 pm	4 pm	3.20 pm

⑦ What is the missing number in this equation?

$12^2 - 9^2 = ? \times 1000$

A	B	C	D	E
63	144	6.3	0.63	0.063

Questions continue on next page

8 Mark created the table below to show the cost of buying electrical items for his new room.

What value should replace X in the table?

Item	Price per item	Number required	Total price
Switches	£2.25	4	£10
Wiring cable rolls	£8.50	2	£17
Radiators	£55	1	£55
Lights	X	3	Y
Total price			£136

A	B	C	D	E
£12	£54	£18	£82	£11

9 If the number 21 is divided by 1000, what is the place value of the number 1 in the answer?

A	B	C	D	E
1 hundred	1 thousand	1 thousandth	1 hundredth	1 tenth

10 The shape below is a kite. Which statement about this shape is TRUE?

A	The shorter sides are not equal in length.
B	The opposite sides are equal in length.
C	The sum of the interior angles is 180°.
D	It has no pairs of parallel lines.
E	It has no line of symmetry.

Score: .... / 10

Test 8

You have 10 minutes to complete this test.

You have 10 questions to complete within the given time.

Circle the letter next to the correct answer.

① Which of the following is equivalent to 2^7?

A	B	C	D	E
2×7	7×7	$2 \times 2 \times 2 \times 2 \times 2 \times 2 \times 2$	7×2	$2 + 2 + 2 + 2 + 2 + 2 + 2$

② Below is a regular pentagon with an area of B cm^2.

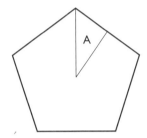

Which of the following is the best estimation of the area of Triangle A?

A	B	C	D	E
$\frac{B}{5}$ cm^2	$5B$ cm^2	$\frac{B}{2}$ cm^2	$\frac{B}{10}$ cm^2	$10B$ cm^2

③ What is the highest common factor of 80, 12 and 20?

A	B	C	D	E
4	24	16	2	8

④ A number is doubled and then 11 is added to it. The final answer is 22.

What is the original number?

A	B	C	D	E
11.5	5.5	11	4.5	6.5

Questions continue on next page

(5) Two consecutive odd numbers add up to 32.

What is the greater of the two numbers?

A	B	C	D	E
14	15	16	17	18

(6) What is the perimeter in centimetres of the larger rectangle in the diagram below?

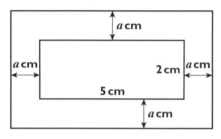

A	B	C	D	E
12 + 4a	10 + 2a	7 + 4a	14 + 8a	7 + 8a

(7) Which of the following ratios is equivalent to 15:2.5?

A	B	C	D	E
1:6	6:1	3:5	5:3	15:2

(8) Which number is in the wrong place in the diagram below?

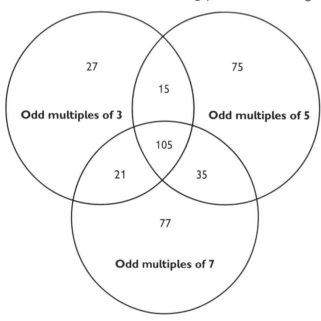

A	B	C	D	E
15	77	105	27	75

9 Which of these shapes has a rotational symmetry of 2?

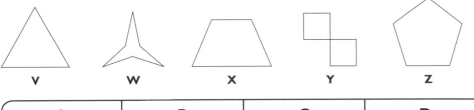

A	B	C	D	E
V	Z	X	Y	W

10 The nutrition value of 100 ml of a fresh juice is shown in the table below.

Carbohydrates	80%
Fat	4%
Protein	5%
Other nutrients	11%

Gina creates a pie chart using the information in the table.

What is the angle at the centre of the pie chart in the slice that represents protein?

A	B	C	D	E
5°	36°	18°	80°	180°

Test 9

You have 10 minutes to complete this test.

You have 10 questions to complete within the given time.

Circle the letter next to the correct answer.

1. Which statement below is TRUE concerning 5.52 and 0.552?

A	One number is double the other number.
B	One number is a tenth of the other number.
C	One number is a hundredth of the other number.
D	One number is thousand times smaller than the other number.
E	One number is a hundred times larger than the other number.

2. The diagram of a bench below has a scale of 1:55.

3 cm

What is the actual length of the bench rounded to the nearest metre?

A	B	C	D	E
1 m	2 m	16 m	1.6 m	3 m

3. The sum of Emily's age and her brother's age is 21. Emily is 5 years younger than her brother.

How old is Emily's brother?

A	B	C	D	E
12	11	8	13	9

④ Figure B is formed from a square and a triangle. What is the area of Figure B?

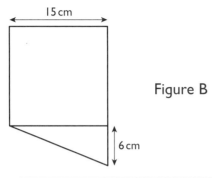

Figure B

A	B	C	D	E
270 cm²	225 cm²	315 cm²	90 cm²	800 cm²

⑤ What is the reading shown by the arrow on the scale below?

A	B	C	D	E
0.17	0.14	0.13	0.3	1.3

⑥ Mrs Brown can sew 7 dresses in five days.

At the same speed, how many weeks does she need to sew 49 dresses if she works everyday?

A	B	C	D	E
35	5	8	12	3

⑦ The table below shows the output of a function machine for given inputs.

Input	5	8	10	15	?
Output	33	54	68	103	138

Using this function machine, what input will give an output of 138?

A	B	C	D	E
14	200	13	20	17

Questions continue on next page

8 Graham's coin collection consists of identical coins with a total weight of 176 g.

If each coin has a weight of 4.4 g, how many coins does Graham have in his collection?

A	B	C	D	E
20	30	40	50	60

9 Fred uses identical small cubes to build three larger cubes that together form Figure D.

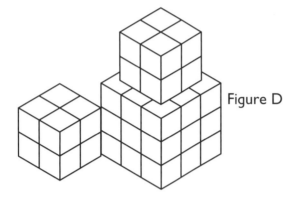

Figure D

What is the total volume of Figure D if the length of each small cube is 3 cm?

A	B	C	D	E
27 cm³	43 cm³	64 cm³	81 cm³	1161 cm³

10 Which of these shapes has no lines of symmetry?

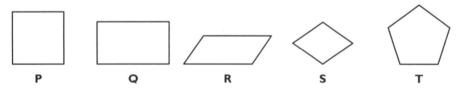

P   Q   R   S   T

A	B	C	D	E
S	T	R	P	Q

Score: .... / 10

Test 10

You have 10 minutes to complete this test.

You have 10 questions to complete within the given time.

Circle the letter next to the correct answer.

(1) Which of the following does not have a vertical line of symmetry?

A	B	C	D	E
AUA	BIB	WOW	MOM	VAV

(2) Ben asked 48 children to choose their favourite chocolate. He recorded his results in the table below.

	Dark chocolate	Milk chocolate	Nut chocolate
Boys	3	7	8
Girls	?	12	17

How many girls chose dark chocolate?

A	B	C	D	E
3	2	16	1	4

(3) A large plant requires 3.5 litres of water a day.

How many litres of water does it require in the month of April?

A	B	C	D	E
105 litres	110 litres	35 litres	70 litres	115 litres

(4) Which of the following is not a cube number?

A	B	C	D	E
27	64	216	161	8

Questions continue on next page

31

(5) The numbers in each row and column in the magic square below must add up to 12.

Which number should replace the question mark?

7	2	
?		
	6	I

A	B	C	D	E
9	8	3	4	0

(6) What is the probability that a randomly chosen day of the week has exactly 6 letters in its spelling?

A	B	C	D	E
$\frac{3}{7}$	$\frac{5}{7}$	$\frac{4}{7}$	$\frac{1}{2}$	$\frac{1}{3}$

(7) Which of these house names has a ratio of vowels to consonants of 1:1?

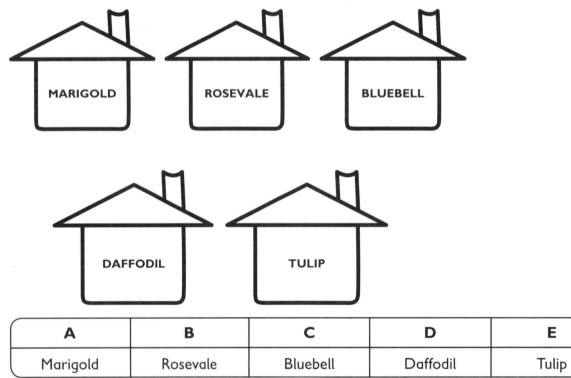

A	B	C	D	E
Marigold	Rosevale	Bluebell	Daffodil	Tulip

(8) The table below shows the weight of five children in Year 6.

Find the range of their weights.

| Weight | 29.5 kg | 33.5 kg | 29.7 kg | 41.3 kg | 36 kg |

A	B	C	D	E
11.8 kg	11.6 kg	6.3 kg	4 g	11 kg

(9) Thomas receives £6.50 pocket money per week.

For how many weeks must he save his pocket money to be able to buy a scooter costing £52?

A	B	C	D	E
7	8	10	4	2

(10) The height of a wall is 973 cm.

What is the height of the wall rounded to the nearest metre?

A	B	C	D	E
1 m	10 m	8 m	97 m	9 m

Score: .... / 10

Test 11

You have 10 minutes to complete this test.

You have 10 questions to complete within the given time.

Circle the letter next to the correct answer.

① A rope is 5.07 m long. It is cut into 3 equal pieces. Each of these 3 pieces is again cut in half.

What is the length of each piece of rope?

A	B	C	D	E
845 cm	8.45 cm	84.5 mm	845 mm	8.45 mm

② The table below shows the performance of the Holdean cricket club over the spring and summer seasons.

Season	Matches won	Matches lost	Matches drawn
Spring	12	3	5
Summer	7	6	7

How many matches in total did Holdean cricket club not win?

A	B	C	D	E
20	9	12	19	21

③ The train ticket to Broadstairs beach costs £24.50 per person. 25 tourists go to the beach by train.

How much money do the tourists pay for their train tickets in total?

A	B	C	D	E
£610.50	£612.50	£586	£600.50	£608

④ A book has 520 pages. 25% of the pages in the book contain illustrations.

How many pages in the book do not contain illustrations?

A	B	C	D	E
39	130	390	200	170

(5) The graph below shows the location of four places:

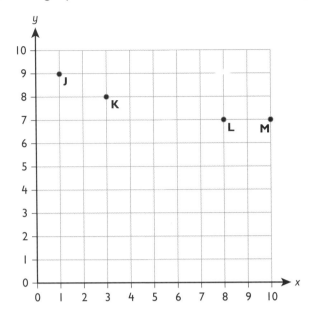

J: Will's house

K: The lake

L: Will's school

M: Tennis court

Which place is shown at the coordinate point (8 , 7)?

A	B	C	D	E
Will's house	Will's school	The lake	Tennis court	None of the above

(6) The water jug below holds enough water to fill 8 glasses.

Which of the following is the best estimate for the total capacity of the water jug?

A	B	C	D	E
1.75 ml	17.5 litres	1.75 litres	20 ml	220 litres

(7) Ronnie ate $\frac{2}{7}$ of his chocolate buttons and left the remaining 15 for his sister.

How many chocolate buttons were there in total?

A	B	C	D	E
20	15	6	21	27

Questions continue on next page

8 In a decade's time, Jean's age will be a third of her father's age.

If her father's current age is b, how old is Jean now?

A	B	C	D	E
$\frac{b + 10}{3} - 10$	$\frac{b + 10}{3} + 10$	$\frac{b - 10}{3} - 10$	$b + 10$	$b - 10$

9 The function machine below divides by 7 and then adds 10.

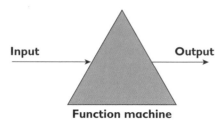

Input → Output

Function machine

Input	1085
Output	?

What is the output for the given input?

A	B	C	D	E
145	155	165	200	175

10 The table below shows the number of cars and the number of passengers in each car that passed by Rope Street one afternoon.

Number of people in the car	5	4	3	2	1
Number of cars	10	12	3	20	30

If a pie chart is drawn to represent the number of cars, what angle at the centre of the pie chart will represent the number of cars with two passengers?

A	B	C	D	E
12°	45°	96°	144°	180°

Test 12

You have 10 minutes to complete this test.

You have 10 questions to complete within the given time.

Circle the letter next to the correct answer.

1 A squash is made by mixing mango juice and water at a ratio of 1:7.

How much water is there in 2.5 litres of squash?

A	B	C	D	E
200 ml	2200 ml	312.5 ml	285.5 ml	2187.5 ml

2 Each time a ping-pong ball bounces, it reaches a height equivalent to 60% of its height on the previous bounce. The table below shows the height reached by the ping-pong ball after each bounce.

Which number should replace the question mark in the table?

Bounce	Height
1	500 cm
2	300 cm
3	180 cm
4	?

A	B	C	D	E
200 cm	90 cm	120 cm	119 cm	108 cm

3 What is the area of the rectangular carpet below?

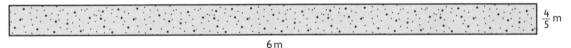

$\frac{4}{5}$ m

6 m

A	B	C	D	E
48 cm²	48000 cm²	4800 cm²	48 cm²	480 cm²

Questions continue on next page

(4) A metal cube, with a side length of 50 cm, is melted to make smaller cuboids each measuring 4 cm by 3 cm by 10 cm.

How many complete cuboids can be made from the metal cube?

A	B	C	D	E
1041	1042	1000	8	100

(5) Figure B consists of a regular pentagon and a triangle.

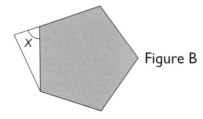

Figure B

What is the size of Angle X?

A	B	C	D	E
60°	20°	70°	72°	10°

(6) What is the value of a in this equation?

$12 - 5a = a - 12$

A	B	C	D	E
3	4	9	2	5

(7) Figure C is a right-angled triangle.

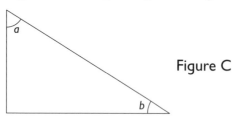

Figure C

Which of these statements about Figure C is TRUE?

A	There are no right angles in Figure C.
B	There is only one acute angle in Figure C.
C	Angle a is obtuse.
D	Angles a and b are both acute.
E	None of the above statements is true.

8. Which of these numbers does not give the value 10.3 when rounded to the nearest tenth?

A	B	C	D	E
10.25	10.34	10.31	10.29	10.35

9. The table below shows the number of children who participate in different activities at school.

Swimming	Fencing	Tennis	Basketball	Choir
24	11	30	?	13

Basketball is the most popular activity.

If the range of the number of children participating in the different activities is 22, how many children play basketball?

A	B	C	D	E
33	14	32	22	10

10. Below is part of the timetable at Brighton Railway Station.

14:32	⊙	**Brighton (Plat 3)**	16 m
14:48	○	**Haywards Heath (Plat 3)**	
14:35	⊙	**Brighton (Plat 6)**	20 m
14:55	○	**Haywards Heath (Plat 3)**	
14:43	⊙	**Brighton (Plat 2)**	23 m
15:06	○	**Haywards Heath (Plat 4)**	

Ali reaches Brighton Railway Station at 14:36.

Assuming all trains are running on time, what is the earliest time he can reach Haywards Heath?

A	B	C	D	E
14:55	15:06	14:35	14:48	14:50

Test 13

You have 10 minutes to complete this test.

You have 10 questions to complete within the given time.

Circle the letter next to the correct answer.

① A piggy bank contains only 1 pence and 2 pence coins. There are three 2 pence coins for every seven 1 pence coins.

If there is a total of £2.60 in the piggy bank, how many 2 pence coins are there?

A	B	C	D	E
120	70	13	140	60

② A regular pentagon and a right-angled triangle are shown below.

What is the sum of the angles A and B?

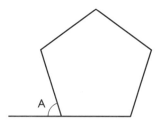 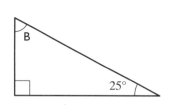

A	B	C	D	E
72°	137°	173°	65°	108°

③ What type of numbers are shown below?

1, 3, 6, 10, 15, 21, 28, 36….

A	Prime numbers
B	Square numbers
C	Cube numbers
D	Triangular numbers
E	Consecutive numbers

(4) A worker is paid 10p per lid for the first 300 toothpaste lids that he screws on. He is then paid 50% more for every additional lid that he screws on.

The table below shows the number of lids he screwed on over 4 separate days.

Day	Monday	Wednesday	Friday	Sunday
Lids screwed	200	400	130	100

How much was the worker paid on Wednesday?

A	B	C	D	E
£40	£45	£400	£10	£100

(5) A recipe for apple crumble uses these ingredients:

Apples	625 g
Butter	135 g
Caster sugar	25 g
Flour	275 g
Oats	30 g
Almonds	50 g

What do the scales read, in kg, when all of these ingredients are mixed together?

A	B	C	D	E
1140	140	1.14	114	104

(6) The table below shows the number of children participating in different clubs at school.

Chess	Scrabble	Jujitsu	Drama	Orchestra
24	11	30	?	13

If the mean number of children participating in each club is 18, how many children participate in Drama?

A	B	C	D	E
18	14	12	20	10

Questions continue on next page

(7) The shape below is reflected in the mirror line.

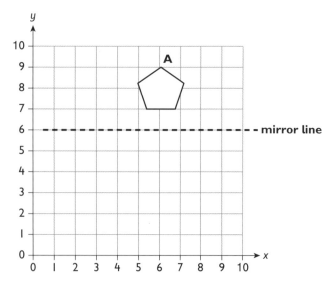

What are the coordinates of the reflection of Point A?

A	B	C	D	E
(3 , 6)	(6 , 2)	(6 , 4)	(4 , 6)	(6 , 3)

(8) A number rounded to the nearest tenth is 51.3

Which of these could be the original number?

A	B	C	D	E
52.34	51.21	51.29	51.35	51.53

(9) What is the value of n in this equation?

$12n - 5 = 9n + 28$

A	B	C	D	E
13	11	9	20	15

(10) What is the acute angle between the hands of a clock at 7.30 pm?

A	B	C	D	E
15°	45°	30°	135°	180°

Score: .... / 10

Test 14

You have 10 minutes to complete this test.

You have 10 questions to complete within the given time.

Circle the letter next to the correct answer.

① Which of these best describes the time 23:50 on a 24-hour clock?

A	Just after midnight
B	Just before lunch
C	Just before midnight
D	Just before 7 pm
E	Almost midday

② What weight should be placed on the left hand side of the scales below, so that the weights on both sides are balanced?

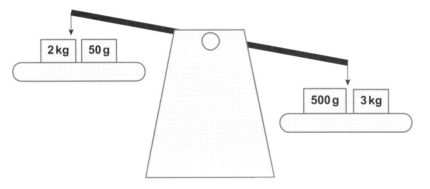

A	B	C	D	E
1.450 g	1450 kg	1 kg	1.45 kg	2300 g

③ Robin uses a calculator to subtract 119.7 from 347 and then adds 456.

What answer will the calculator display?

A	B	C	D	E
683.3	677.3	618.3	567.3	638.3

Questions continue on next page

(4) Isabel chooses copies of the 2D shapes below to assemble a 3D triangular prism.

How many copies of each 2D shape will she need to make the 3D triangular prism?

	Shape A	Shape B

A	2 of Shape A and 2 of Shape B
B	3 of Shape A and 2 of Shape B
C	4 of Shape A and 2 of Shape B
D	1 of Shape A and 2 of Shape B
E	1 of Shape A and 3 of Shape B

(5) Two different odd numbers between 1 and 10 add up to 12.

Which of these statements is TRUE?

A	One of the numbers could be 1.
B	One of the numbers could be 6.
C	The greater of the two numbers must be less than 8.
D	9 could be one of the numbers.
E	No two odd numbers less than 10 add up to 12.

(6) What is the answer to the sum below?

$(118.78 \div 10) \times (118.78 \div 118.78)$

A	B	C	D	E
1.1878	11.878	118.78	1	0

(7) At Jerry's party, the boys drank 9 litres of cola while the girls drank 6 litres of cola.

What fraction of all the cola did the girls drink?

A	B	C	D	E
$\frac{3}{5}$	$\frac{2}{3}$	$\frac{2}{5}$	$\frac{1}{5}$	$\frac{6}{11}$

(8) What is the smaller angle between the hands of a clock at 1.30 pm?

A	B	C	D	E
135°	45°	30°	105°	90°

(9) The graph below shows the remaining battery charge of a toy over a period of time.

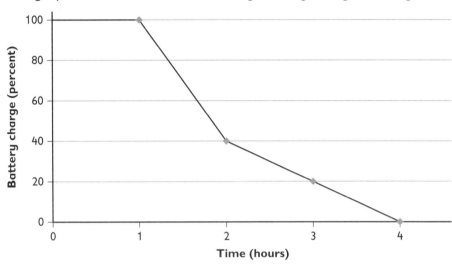

Which of these statements is FALSE?

A	The battery lasts for 4 hours in total.
B	After 1 hour, the battery remains fully charged.
C	After 3 hours, the battery is 50% charged.
D	After 2 hours, the battery is 40% charged.
E	The battery loses more charge in the first 2 hours than the next 2 hours.

(10) The table below shows the number of plants sold by a garden centre over 6 days.

	Monday	Tuesday	Wednesday	Thursday	Friday	Saturday
Number of plants sold	112	98	51	34	160	197

What is the median number of plants sold by the garden centre?

A	B	C	D	E
35	105	57	98	108

Score: .... / 10

Test 15

You have 10 minutes to complete this test.

You have 10 questions to complete within the given time.

Circle the letter next to the correct answer.

(1) Mrs Lehmann places 243 erasers into packs of 30.

How many erasers are left over?

A	B	C	D	E
2	1	12	3	8

(2) Which of these tile shapes will not fit together with identical copies of themselves to form a tile pattern without leaving any gaps?

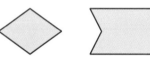

V	W	X	Y	Z

A	B	C	D	E
W	Z	X	Y	V

(3) Which pair of numbers is equally distant from 15?

A	B	C	D	E
14.88 15.12	14.88 15.22	14.8 15.8	13.5 17.5	11.3 15.3

(4) Which expression represents the value of the number marked by the question mark on the number line below?

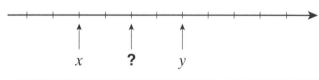

x **?** y

A	B	C	D	E
$x + y$	$x + \frac{y}{2}$	$(x + y) \div 2$	$2xy$	$y - x$

(5) One 10 pence coin weighs 6.5 grams. Alfred has £8 worth of 10 pence coins.

What is the total weight of Alfred's coins?

A	B	C	D	E
0.52 g	0.52 kg	52 g	502 g	520 kg

(6) What are the coordinates of the reflection of Point A in the mirror line?

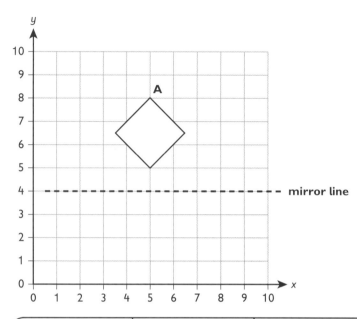

A	B	C	D	E
(0 , 5)	(5 , 3)	(3 , 5)	(5 , 0)	(3.5 , 6.5)

(7) Derek thinks of a 4-digit number.

Which of these statements must be TRUE if Derek's number is a multiple of 8?

A	The sum of all the digits is 8.
B	The last three digits are divisible by 16.
C	The last three digits are divisible by 8.
D	All the digits are even.
E	The number is a multiple of 16.

Questions continue on next page

(8) Below is a diagram of a rectangular park which consists of a tennis court that is surrounded by a lawn 3 metres wide.

What is the outer perimeter of the rectangular park?

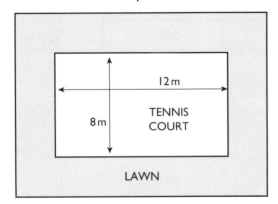

A	B	C	D	E
32 m	40 m	20 m	64 m	96 m

(9) $16 > x^2$

If x is a whole number, which of these is the largest possible value of x?

A	B	C	D	E
3	4	1	5	6

(10) The map below shows a number of different roads. Which two roads are perpendicular to each other?

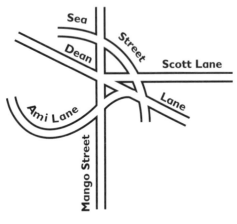

A	Mango Street and Dean Lane
B	Ami Lane and Scott Lane
C	Mango Street and Scott Lane
D	Dean Lane and Ami Lane
E	None of the above

Score: .... / 10

Test 16

You have 10 minutes to complete this test.

You have 10 questions to complete within the given time.

Circle the letter next to the correct answer.

1 Mrs Baker is 5 feet 4 inches tall. Which is closest to her height in metres?

A	B	C	D	E
1.6 m	1.8 m	1.7 m	1 m	2 m

2 Matchsticks are arranged in squares in a sequence as shown below.

Sam writes down the number of small squares in each arrangement.

What type of numbers does Sam write down?

A	B	C	D	E
Decimal fractions	Square numbers	Cube numbers	Triangular numbers	Natural numbers

3 What fraction of the days of the week are spelled with more than 7 letters?

A	B	C	D	E
$\frac{4}{7}$	0.2857	0.528	$\frac{1}{7}$	$\frac{3}{7}$

4 Ellie draws a plan of her house using a scale of 3 cm to 7 m. On the plan, her living room is 5.1 cm long.

What is the actual length of her living room?

A	B	C	D	E
11.9 m	15.3 m	21 m	35.7 m	7 m

Questions continue on next page

(5)

Wendy guides her toy bicycle along the grey squares on the grid to the right.

The bicycle starts on the square marked 'START' and finishes on the square marked 'STOP'.

The bicycle can only move forward, turn left 90° and turn right 90°.

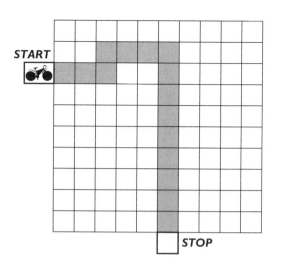

Which instructions will guide the toy bicycle along the grey squares on the grid?

A	FORWARD 3, TURN LEFT 90°, FORWARD 1, TURN RIGHT 90°, FORWARD 3, TURN RIGHT 90°, FORWARD 9
B	FORWARD 2, TURN LEFT 90°, FORWARD 1, TURN RIGHT 90°, FORWARD 3, TURN RIGHT 90°, FORWARD 9
C	FORWARD 3, TURN LEFT 90°, FORWARD 2, TURN RIGHT 90°, FORWARD 3, TURN RIGHT 90°, FORWARD 9
D	FORWARD 3, TURN RIGHT 90°, FORWARD 1, TURN RIGHT 90°, FORWARD 3, TURN RIGHT 90°, FORWARD 9
E	FORWARD 3, TURN LEFT 90°, FORWARD 1, TURN RIGHT 90°, FORWARD 3, TURN RIGHT 90°, FORWARD 11

(6) This graph shows the relationship between weights in pounds and kilograms:

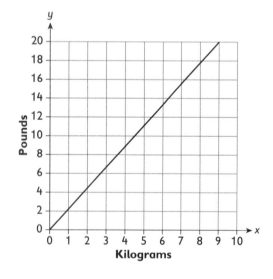

Lucy weighs 81 kg. Which of the following is the closest approximation of Lucy's weight in pounds?

A	B	C	D	E
190	110	20	180	36

7 A childminder works for 7 hours per day and 5 days per week. He earns £17 an hour.

How much does he earn in 4 weeks?

A	B	C	D	E
£595	£2408	£2380	£35	£850

8 How many of the shapes below have more than 1 line of symmetry?

A	B	C	D	E
0	2	1	3	4

9 The cost in pence of 12 erasers is given by the expression $3a - 1$.

What is the cost of each eraser in pounds?

A	B	C	D	E
$\dfrac{(3a - 1)}{1200}$	$(3a - 1) \times 12$	$\dfrac{12 \times (3a - 1)}{100}$	$12 \times (3a - 1) \times 100$	$36a$

10 This pattern is formed from identical regular hexagons.

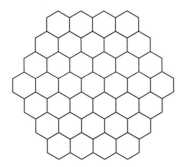

Which of these statements is TRUE?

A	The shape shows a tessellation of regular hexagons.
B	Regular hexagons can never tessellate.
C	No regular polygons can tessellate.
D	Octagons can tessellate.
E	None of the above

Score: .... / 10

51

Test 17

You have 10 minutes to complete this test.

You have 10 questions to complete within the given time.

Circle the letter next to the correct answer.

① A box contains a mixture of red and green peppers. There are 45 peppers in total. There are 11 more green peppers than red peppers.

How many green peppers are there in the box?

A	B	C	D	E
34	11	17	28	29

② A plane is currently flying from Dubai to London. The plane's departure was delayed. The flight details are shown below:

Dubai, Terminal 3	London Gatwick, North
14:55	**19:45**
Thu 27 Apr	
Scheduled departure	Scheduled arrival
15:32	**20:08**
Thu 27 Apr	
Actual departure	Estimated arrival

How many minutes late is the flight's arrival estimated to be?

A	B	C	D	E
37 minutes	23 minutes	16 minutes	8 minutes	45 minutes

③ Katie's café has 19 tables, each with a container for salt. Each container holds 120 g of salt.

How much salt does she need to fill all the 19 containers from empty?

A	B	C	D	E
2 kg	2.2 kg	2.28 kg	3.26 kg	3.7 kg

4 Which of these letter sequences has a vertical line of symmetry?

A	B	C	D	E
EVE	ASA	MUM	LOL	WCW

5 What is the number below, rounded to 3 decimal places?

76.23958

A	B	C	D	E
76.2	76.239	76.240	76.2396	76.23

6 The measurements of two sides of a rectangle are shown below.

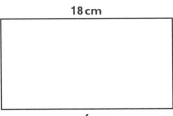

18 cm

$x - 6$ cm

What is the value of x?

A	B	C	D	E
12 cm	24 cm	108 cm	180 cm	6 cm

7 Figure D is a quadrilateral made from an equilateral triangle and a parallelogram.

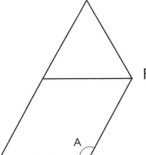

Figure D

What is the measurement of Angle A?

A	B	C	D	E
60°	120°	180°	160°	10°

Questions continue on next page

(8) Ewan's scores in 11 games of Scrabble are shown below:

211   110   96   123   35   167   132   111   110   333   49

What is his median score?

A	B	C	D	E
123	111	132	110	49

(9) What transformation must be made to Figure E so that it looks like Figure F?

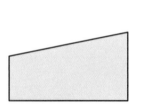

Figure E   Figure F

A	B	C	D	E
Division	Rotation	Reflection	Enlargement	None of the above

(10) The price of a guitar was reduced by 20%. Sandra paid the reduced price of £120.

How much money did Sandra save on her guitar purchase?

A	B	C	D	E
£150	£20	£30	£120	£40

Score: .... / 10

Test 18

You have 10 minutes to complete this test.

You have 10 questions to complete within the given time.

Circle the letter next to the correct answer.

1 Students are tested in 8 subjects in their exam paper. Each subject consists of 5 sections, each with 16 questions.

How many questions are there in the exam paper in total?

A	B	C	D	E
400	800	640	90	40

2 The diagram below shows the floor plan of Sam's garage.

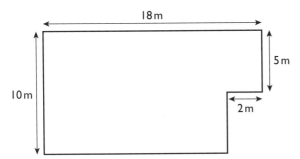

What is the area of Sam's garage?

A	B	C	D	E
180 m²	110 m²	90 m²	170 m²	10 m²

3 A maximum of 48 tissue rolls fit into a case. A worker packs 900 tissue rolls into cases.

How many full cases are packed?

A	B	C	D	E
36	18	32	19	36

Questions continue on next page

4 Below is a regular pentagon, with Point F at its centre.

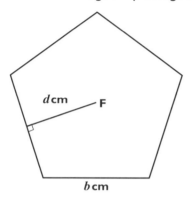

Which expression shows the area of the pentagon?

A	B	C	D	E
bd cm²	$\frac{bd}{2}$ cm²	$2.5\ bd$ cm²	$5\ bd$ cm²	$\frac{d}{2}$ cm²

5 Which expression is equivalent to $(2^2)^3$?

A	B	C	D	E
3 × 3 × 3	2 × 2 × 2	3 × 2 × 4	4 × 4 × 4	8 × 2

6 Which number should replace the question mark to complete the grid below?

3.25	3.80	4.35
3.80	4.35	4.9
4.35	4.9	?

A	B	C	D	E
5.25	5.45	5.15	5.05	4.95

7 Which ratio is equivalent to 104:143?

A	B	C	D	E
52:71	8:13	8:17	8:11	7:11

(8) The shape below consists of an isosceles triangle and a rectangle.

What is the value of Angle *X*?

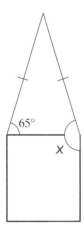

65°

X

A	B	C	D	E
65°	135°	155°	90°	45°

(9) The probability that a team wins a football match is 0.62. There is a 0.3 probability that the match will end in a draw.

What is the probability that the team loses the match?

A	B	C	D	E
0.35	0.80	0.08	0.12	0.92

(10) Figure E is formed from identical cubes.

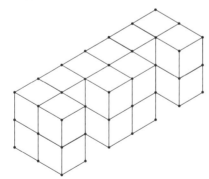

Figure E

What is the volume of Figure E if each cube has a side length of 3 cm?

A	B	C	D	E
540 cm³	270 cm³	500 cm³	320 cm³	27 cm³

Score: .... / 10

Test 19

You have 10 minutes to complete this test.

You have 10 questions to complete within the given time.

Circle the letter next to the correct answer.

(1) Which of these numbers is not a multiple of 9?

A	B	C	D	E
1044	2808	1649	909	4806

(2) How many of the shapes below have at least one internal reflex angle?

    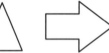

A	B	C	D	E
1	2	3	4	0

(3) What is the ratio 49:343, reduced to its lowest terms?

A	B	C	D	E
7:49	49:7	1:49	49:49	1:7

(4) The table below shows the bowling scores of 4 friends.

Jo	Benson	Cathy	Derek
78	49	34	90

What is the range of the scores?

A	B	C	D	E
56	41	44	90	22

(5) Which two lines on the grid below are parallel to each other?

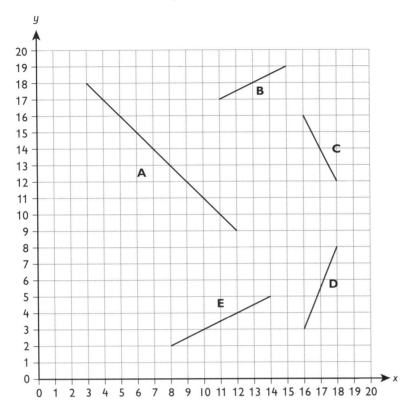

A	B	C	D	E
A and B	B and D	B and E	A and C	D and E

(6) In a survey of 132 people, every fourth person had a trampoline at home.

How many people had a trampoline at home?

A	B	C	D	E
0.4 × 132	4 × 132	0.25 × 132	$132 - \frac{1}{4}$	$132 + \frac{1}{4}$

(7) Each term in the sequence below is found by multiplying the previous term by 5 and then adding 1.

?, 7, 36, 181,.....

Which number should replace the question mark in the sequence?

A	B	C	D	E
6	1.2	1.1	35	34

Questions continue on next page

(8) Fred asked his friends what their preferred drink was and then created the pie chart below to present their responses.

Which of these statements is TRUE?

A	More friends preferred tea than preferred milk.
B	The angle representing the number of friends who preferred milk is acute.
C	Roughly a quarter of the friends preferred coffee.
D	More than half the friends preferred hot chocolate.
E	Roughly three-quarters of the friends preferred milk.

(9) Grace cuts out 2 identical triangles and 1 rectangle from a piece of card to create Figure A.

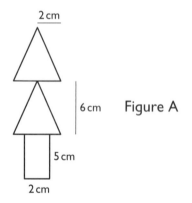

Figure A

What is the area of Figure A?

A	B	C	D	E
24 cm²	32 cm²	34 cm²	22 cm²	60 cm²

(10) Which of the following has the second greatest value?

A	B	C	D	E
101	50% of 205	50.75 × 2	$\frac{1}{3}$ of 312	0.25 of 400

Score: .... / 10

Test 20

You have 10 minutes to complete this test.

You have 10 questions to complete within the given time.

Circle the letter next to the correct answer.

(1) Eddie calculates the answer to 199 × 307.

Which of the following is closest to his answer?

A	B	C	D	E
300000	60000	600	600000	3000

(2) The chart below shows the scores, as a percentage, attained by all the students in a group.

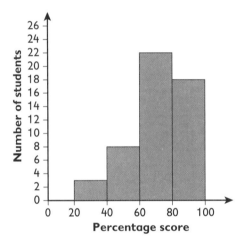

How many students are in the group?

A	B	C	D	E
37	51	33	53	45

(3) What is the volume of a cube with sides of $2x$ cm?

A	B	C	D	E
$2x^3$ cm³	$8x$ cm³	$8x^3$ cm³	$2x$ cm³	$16x^3$ cm³

Questions continue on next page

4 Which of these letter sequences has a horizontal line of symmetry?

A	B	C	D	E
ETE	XOX	POP	WIM	OPO

5 An equilateral triangle and a parallelogram with the same height are shown below.

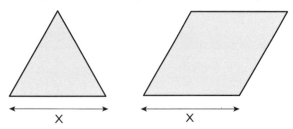

Which of these statements is TRUE?

A	The areas of the 2 shapes are equal.
B	The area of the triangle is double that of the parallelogram.
C	The area of the triangle is one-third that of the parallelogram.
D	The area of the triangle is half that of the parallelogram.
E	The area of the triangle is 70% of the area of the parallelogram.

6 Below is a square with the dimensions shown.

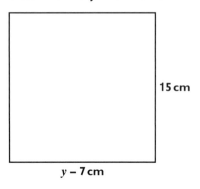

What is the value of y?

A	B	C	D	E
22 cm	8 cm	105 cm	18 cm	15 cm

7 $j \div (5 \times 4) = 1.5$

What is the value of j?

A	B	C	D	E
300	20	15	30	3

(8) The diagram below shows a regular pentagon on a centimetre grid.

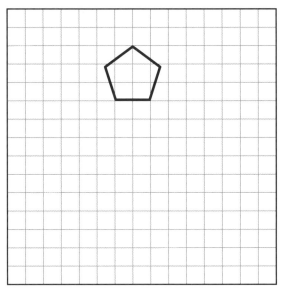

If the shape is enlarged by a scale factor of 3, how long is each enlarged side of the pentagon?

A	B	C	D	E
3 cm	5 cm	15 cm	6 cm	12 cm

(9) A milkman delivers 1.5 litres of milk to 80 houses per day. He delivers milk every day except Sunday.

How much milk does he deliver in a week?

A	B	C	D	E
72 litres	720 litres	7200 litres	8200 litres	None of the above

(10) The table below shows the classification of different objects in a lab.

	Metal	Non-metal
Red	29	1
Blue	30	14

How many more metal objects are there in the lab than blue non-metal objects?

A	B	C	D	E
1	15	16	24	45

Score: .... / 10

Test 21

You have 10 minutes to complete this test.

You have 10 questions to complete within the given time.

Circle the letter next to the correct answer.

(1) Out of 576 people who registered for a charity walk, 75% completed the walk.

How many of those registered did not complete the walk?

A	B	C	D	E
143	144	212	432	300

(2) Figure A is a large cube made up of identical smaller cubes.

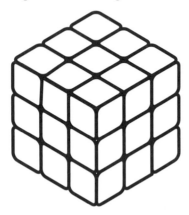

Figure A

Ruby draws a blue star on two-thirds of the smaller cube faces.

How many of the smaller cube faces have a star drawn on them?

A	B	C	D	E
54	27	9	18	36

(3) X represents a number. When X is rounded to the nearest whole number, the answer is 50. When X is rounded to the nearest tenth, the answer is 49.6.

Which of the following could be X?

A	B	C	D	E
51.56	49.51	49.58	50.1	49.85

④ Below is an architect's diagram of a garden. The garden will consist of a square lawn and a water feature. The scale of the diagram is 1:120.

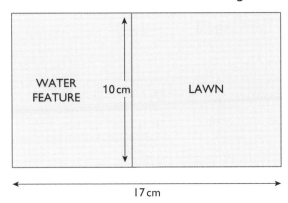

What is the actual area of the land allotted for the water feature?

A	B	C	D	E
1008 m²	100.8 m²	10.08 m²	1.008 m²	10 m²

⑤ Laila groups 624 cards into equal piles of 6. How many piles does Laila make?

A	B	C	D	E
103	14	104	156	68

⑥ In the diagram below, each number is the product of the two numbers directly below it.

80	200	600	1560

20	4	50	12	?

Which number should replace the question mark?

A	B	C	D	E
130	140	120	180	15

⑦ A machine uses 50 cm of wool to create a hat.

If the machine makes 12 hats every day, how much wool does it use in 1 week?

A	B	C	D	E
42 cm	4200 m	420 m	42 m	420 cm

Questions continue on next page

(8) Below is a conversion table for metres to feet.

Metres	1	3	5	6	?
Feet	3.28	9.84	16.4	19.68	29.52

Which number should replace the question mark?

A	B	C	D	E
3	15	10	9	12

(9) Stella drinks $\frac{3}{5}$ of a full carton of water. The carton has a capacity of 2.5 litres.

How much water is left in the carton?

A	B	C	D	E
1 litre	720 ml	1.5 litres	1200 ml	2 litres

(10) The table below shows the classification of 48 plants in a garden.

	Shrub	Climber
Hardy	13	10
Annual	11	?

How many of the plants are Annual Climbers?

A	B	C	D	E
17	14	11	13	12

Score: .... / 10

Test 22

You have 10 minutes to complete this test.

You have 10 questions to complete within the given time.

Circle the letter next to the correct answer.

1. The table below shows the power consumption per day of different types of lights.

Item	Power consumed in watts per day
10 tree lights	48
15 rope lights	120
90 bulb lights	300

Zara has 10 tree lights, 45 rope lights and 30 bulb lights.

How much power do Zara's lights consume in 2 days?

A	B	C	D	E
408 watts	508 watts	330 watts	1016 watts	1018 watts

2.

22	29	36	43	50
18	24	30	36	42
14	19	24	29	34
10			22	26
6			15	18

Which of the following should replace the empty spaces in the number grid above?

A	B	C	D	E
14 18 / 9 11	14 18 / 9 12	12 18 / 9 11	14 18 / 9 1	14 20 / 9 11

Questions continue on next page

(3) 9 metres of ribbon are cut into b equal pieces. Each of these pieces is then cut in half.

What is the length of each piece of ribbon?

A	B	C	D	E
4.5b cm	$\frac{4.5}{b}$ cm	$\frac{450}{b}$ m	$\frac{450}{b}$ cm	$\frac{9}{2b}$ cm

(4) The number of edges in a square-based pyramid is given by e while the number of vertices is given by v.

Which of the following expressions is correct?

A	B	C	D	E
$v = e + 3$	$e = \frac{v}{2}$	$v = \frac{e}{2}$	$v = e - 3$	$v + e = 10$

(5) What is the reflex angle between the hands on the clock face below?

A	B	C	D	E
130°	190°	210°	240°	150°

(6) Andy, Brian and Callum shared £151.20 equally between them. Callum spent £12.50 of his share on a board game.

How much of Callum's share remained?

A	B	C	D	E
£37.90	£38.35	£33.90	£42.15	£37.10

(7) What is the value of n in the equation below?

$6n - 9 = 165$

A	B	C	D	E
51	27	29	18	31

8 Lakeland Hotel offered a 30% discount on all room bookings and an additional 5% discount for rooms booked before 30th May 2017. Ralph booked a room on 21st April 2017. The original cost of the room was £90.

What was the cost after the discount?

A	B	C	D	E
£48.50	£60	£72.50	£80	£58.50

9 The table below converts shoe size measurements in different regions.

International shoe size conversion			
USA sizes from toddler 12 to adult 12			
US & Canada	Europe	Japan	UK
12	31	175	11
13	32	180	12
1	33	190	13
2	34	200	1
3	35	210	2
4	36	218	3
5	37	221	4
6	38	225	5
7	39	235	6
8	40	250	7
9	41	260	8
10	42	270	9
11	43	280	10
12	44	295	11

Harriet wears size 2 shoes in the UK. What is Harriet's shoe size in Japan?

A	B	C	D	E
35	2	3	221	210

10 The scale below shows Tracey's weight.

45.09 kg

How much does Tracey weigh?

A	B	C	D	E
45 kg and 900 g	45 kg and 90 g	45 kg and 9 g	4 kg and 509 g	45 kg and 99 g

Score: .... / 10

Notes

Answers

Key abbreviations: °C: degrees centigrade, cm: centimetre, g: gram, kg: kilogram, km: kilometre, m: metre, ml: millilitre, mm: millimetre

Test 1

Q1 D

OHO has a horizontal line of symmetry so it will read the same when rotated 180°.

Q2 C

S represents the output value once N has passed through the boxes in the function machine.

1st box → $N \times 2 = 2N$

2nd box → $2N - 1$

3rd box → $4(2N - 1) = 8N - 4$

Q3 C

To form a triangle, the sum of the length of any two sides must be greater or equal to the length of the third side.

Q4 B

Let X be the number of Japanese coins.

Number of Chinese coins = $X + 36$

Total number of coins = $X + X + 36 = 440$

This equation can be solved to find X.

$2X + 36 = 440$

$2X = 404$

$X = 202$

Number of Chinese coins = $X + 36 = 202 + 36 = 238$

Q5 E

The cubes with exactly 3 faces painted blue are marked with an X.

Q6 C

Note that r is expressed in centimetres.

11 m = 1100 cm

Total cord cut off = $8 \times n = 8n$

So $r = 1100 - 8n$

Q7 D

Since the pie chart shows how many times he earned each number of stars, the mode must be represented by the largest slice of the pie chart.

Q8 D

Total number of cakes ordered =

24 + 18 + 12 + 9 + 12 = 75

Mean number of cakes ordered per day = 75 ÷ 5 = 15

Q9 A

$9y - 7 = 29 - 3y$

$9y + 3y = 29 + 7$

$12y = 36$

$y = 3$

Q10 B

Shape F has the same perimeter as a rectangle with length a and width b.

So the perimeter of Shape F = $a + a + b + b = 2a + 2b$

Test 2

Q1 C

30% of 120 = 36

$\frac{1}{2}$ of 80 = 40

Three-fifths of 90 = 54

0.2 of 250 = 50

Q2 C

Remaining amount to pay = £480 − £60 = £420

Number of weeks required to pay full remaining cost

= 420 ÷ 20 = 21

Q3 D

8	3	4
1	5	9
6	7	2

Q4 B

Harry's age now = 10 − 3 = 7

Harry's sister's age now = 7 × 2 = 14

Harry's sister's age in 3 years = 14 + 3 = 17

Q5 B

There are 8 words in total.

5 of them contain more than 4 letters.

Probability = $\frac{5}{8}$

Q6 D

48 ÷ 3 = 16

Smallest number = 16 − 1 = 15

Largest number = 16 + 1 = 17

Check: 15 + 16 + 17 = 48

Q7 B

2 cups of butter = 225 g × 2 = 450 g

3 cups of flour = 150 g × 3 = 450 g

$\frac{1}{5}$ cup of sultanas = 200 g ÷ 5 = 40 g

$\frac{1}{2}$ cup of almonds = 110 g ÷ 2 = 55 g

Total weight = 450 g + 450 g + 40 g + 55 g = 995 g

Q8 C

There are 5 small squares in each row so side length of each small square is 2 cm.

Area of 1 small square = 2 cm × 2 cm = 4 cm²

Number of squares shaded black = 13

Area of Figure C that is shaded black =

13 × 4 cm² = 52 cm²

Q9 B

Figure B consists of 7 identical triangles.

Area of 1 triangle = $\frac{1}{2} \times$ base $\times$ height = $\frac{1}{2} \times 2y \times x$

= $\frac{1}{2} \times 2xy = xy$

Area of Figure B = 7 × xy = $7xy$ square units

Q10 A

40 m = 4000 cm

4000 cm ÷ 2000 = 2 cm

Test 3

Q1 **A**

$1752 + 40 = 1792$

$32 \times 56 = 1792$

Q2 **B**

$300\text{ g} \div 500\text{ g} = \frac{3}{5}$

Amount of coconut milk she should use $= \frac{3}{5}$ of 200 ml

$= 120$ ml

Q3 **A**

An obtuse angle is greater than 90° but less than 180°.

Q4 **C**

Figure C is formed from 3 squares and 2 circles.

Area of Figure C $= 3a + 2b$

Q5 **C**

4 out of 10 shirts are striped so probability $= \frac{4}{10} = \frac{2}{5}$

$\frac{2}{5} = 0.4$

Q6 **D**

Let the length of the rectangle $= X$

$X = (28\text{ cm} - 4\text{ cm} - 4\text{ cm}) \div 2 = 20\text{ cm} \div 2 = 10\text{ cm}$

Area of the rectangle = length × width =

$10\text{ cm} \times 4\text{ cm} = 40\text{ cm}^2$

Q7 **E**

Work backwards:

$1 \times 1000 = 1000$

Cube root of 1000 = 10

Q8 **B**

The most common number of packed lunches is 11 as this occurs 7 times.

Q9 **D**

$11.8\text{ kg} = 11800\text{ g}$

Total weight of the matchboxes $= 11800\text{ g} - 600\text{ g}$

$= 11200\text{ g}$

Weight of one matchbox $= 11200\text{ g} \div 400 = 28\text{ g}$

Q10 **E**

Sixth term $= (70 \times 1045) - 5 = 73150 - 5 = 73145$

Test 4

Q1 **C**

Sum of the other 2 angles $= 180° - 112° = 68°$

Because the triangle is isosceles, 2 angles must be equal. So the value of each of the other angles in the triangle $= 68° \div 2 = 34°$

Q2 **B**

Food packets per child $= x + y$

Total food packets $= 40(x + y)$

Q3 **C**

The graph shows that Andy travels 12 km away from his starting point and then travels a further 12 km to return to his starting point.

Total distance covered $= 12\text{ km} + 12\text{ km} = 24\text{ km}$

Q4 **A**

Weight of remaining cotton $= 331.1\text{ kg} - 119\text{ kg}$

$= 212.1\text{ kg}$

Q5 **C**

Figure A has the equivalent perimeter of a 15 cm by 11 cm rectangle.

Perimeter $= 15\text{ cm} + 15\text{ cm} + 11\text{ cm} + 11\text{ cm} = 52\text{ cm}$

Q6 **E**

9 am on Monday morning to 9 pm on Friday evening is 4.5 whole days.

Number of hours in 1 day = 24

Number of hours in 4.5 days $= 4.5 \times 24 = 108$

Q7 **E**

Total shaded area of Figure B $= 2s$

$\frac{1}{4}$ of Figure B is shaded.

So $\frac{3}{4}$ of Figure B is unshaded.

Unshaded area $= 2s \times 3 = 6s$

Q8 **B**

y must be greater than −11 but less than 5.

Q9 **A**

The shape holds the same form only once in a rotation of 360°.

Q10 **D**

The relationship between x and y is as follows:

$y = x^2 - 1$

When x equals 6: $y = 6^2 - 1 = 36 - 1 = 35$

Test 5

Q1 **C**

26 out of the 52 cards are odd numbered.

$\frac{26}{52} = \frac{1}{2}$

Q2 **E**

The pentagon turns through $\frac{3}{5}$ of a full rotation.

$\frac{3}{5}$ of 360° = 216°

Q3 **E**

$\frac{4}{25}$ does not equal 15%

$\frac{4}{25} = 16\%$

Q4 **D**

Total cost of the two items $= £10 - £4.60 = £5.40$

Cost of Minestrone Soup and Scrambled egg = £3.50 + £1.90 = £5.40

Q5 **C**

$\frac{1}{7}$ of 56 = 8

$\frac{3}{7}$ of 56 $= 8 \times 3 = 24$

Number of stamps that are not vintage $= 56 - 24 = 32$

Number of blue stamps $= \frac{1}{4}$ of 32 = 8

Q6 **D**

Charles must face the opposite direction to south-east, which is north-west.

Q7 **C**

Work backwards:

$36 \times 4 = 144$

Square root of 144 = 12

Q8 **A**

Weight of 4 bananas = 300 g

Weight of 1 banana $= 300\text{ g} \div 4 = 75\text{ g}$

Weight of 6 bananas $= 75\text{ g} \times 6 = 450\text{ g}$

Q9 **D**

Total cost $= £2100 + £700 + £400 = £3200$

Mean cost per person $= £3200 \div 5 = £640$

Q10 **E**

A reflex angle is greater than 180° but less than 360°.

Test 6

Q1 **B**
A = 6 faces, B = 7 faces, C = 4 faces, D = 5 faces, E = 2 faces

Q2 **A**
$-15°C \rightarrow 18°C = 33°C$

Q3 **E**
Total area of lawn = (30 m × 15 m) + (2 m × 2 m) = 450 m² + 4 m² = 454 m²
Cost of mowing lawn = £454x
Cost to water and weed plants = £10
Total cost = £(10 + 454x)

Q4 **A**
Area of four walls = 4 × (4 m × 7 m) = 112 m²
Number of tins required = 112 m² ÷ 12 m² = $9\frac{1}{3}$
Exactly $9\frac{1}{3}$ tins cannot be purchased so the answer is 10 tins.

Q5 **C**
Triangle BCD is isosceles.
Angle BCD = interior angle of a regular hexagon = 120°
So Angle CBD = (180° − 120°) ÷ 2 = 30°

Q6 **B**
Total walking time = $2n + 1 + 3n − 4 + n + 6 + 2n + 5$ = $8n + 8$
Mean time spent walking per day = $(8n + 8) ÷ 4 = 2n + 2$

Q7 **B**
Number of books in Pile 1 = $\frac{2}{5}$ × 210 = 84
Number of books in Pile 2 = $\frac{1}{3}$ × 210 = 70
Number of books in Pile 3 = 10% × 210 = 21
Number of books in Pile 4 = 210 − 84 − 70 − 21 = 35
84 and 21 are multiples of 21.

Q8 **C**
Pocket calculators were invented in 1972, before personal computers.

Q9 **D**
176 is twice as large as 88
So (35728 ÷ 88) will be twice as large as (35728 ÷ 176)
406 ÷ 2 = 203

Q10 **A**
Water drunk by each goat = 96 ÷ 15 = 6.4 l = 6 l (rounded to the nearest litre)

Test 7

Q1 **C**
Triangle-based pyramid and square-based pyramid

Q2 **A**
Flour required for 5 people = 900 g
Flour required for 1 person = 900 g ÷ 5 = 180 g
Flour required for 14 people = 14 × 180 g = 2520 g
2520 g rounded to the nearest kilogram = 3 kg

Q3 **C**
Area of equilateral triangle = $\frac{h}{6}$
Area of Figure B = $3h + \frac{h}{6}$

Q4 **B**
Each term is 6 greater than the previous term. So the 11ᵗʰ term is 65.

Q5 **C**
Percentage of children = 100% − 32% − 41% = 27%
Number of children = 27% × 900 = 243

Q6 **C**
Monday → 30 minutes
Tuesday → 40 minutes
Wednesday → 20 minutes
Thursday → 20 minutes
Time spent playing chess over 5 days = 25 minutes × 5 = 125 minutes
Time spent playing chess on Friday = 125 − 30 − 40 − 20 − 20 = 15 minutes
15 minutes after 3 pm = 3.15 pm

Q7 **E**
$? = \frac{12^2 - 9^2}{1000} = 144 - \frac{81}{1000} = \frac{63}{1000} = 0.063$

Q8 **C**
Y = £136 − £10 − £17 − £55 = £54
X = £54 ÷ 3 = £18

Q9 **C**
21 ÷ 1000 = 0.021
The 1 is in the thousandths column.

Q10 **D**
A kite has no pairs of parallel lines.

Test 8

Q1 **C**
2^7= 2 × 2 × 2 × 2 × 2 × 2 × 2

Q2 **D**
The pentagon can be split into 5 identical triangles. Triangle A appears to be half the area of 1 of these triangles. So Triangle A has an area roughly $\frac{1}{10}$ the size of the pentagon.
Area of Triangle A = B ÷ 10 = $\frac{B}{10}$ cm²

Q3 **A**
4 is the answer choice with the greatest value that is a factor of 80, 12 and 20

Q4 **B**
Work backwards:
22 − 11 = 11
11 ÷ 2 = 5.5

Q5 **D**
15 + 17 = 32

Q6 **D**
Length of rectangle = $5 + 2a$
Width of rectangle = $2 + 2a$
Perimeter = $5 + 2a + 5 + 2a + 2 + 2a + 2 + 2a$ = $14 + 8a$

Q7 **B**
6 × 2.5 = 15; 1 × 2.5 = 2.5
So 6:1 is equivalent to 15:2.5

Q8 **E**
75 is an odd multiple of both 3 and 5 so it should be placed in the intersection of the top two circles.

Q9 **D**
The shape holds the same form twice when rotated through 360°.

Q10 **C**
Total angle at the centre of a pie chart = 360°
Angle that represents protein = 5% × 360° = 18°

Test 9

Q1 **B**
$5.52 \div 10 = 0.552$

Q2 **B**
Length of the bench = 3 cm × 55 = 165 cm = 1.65 m
1.65 m rounded to the nearest metre is 2 m

Q3 **D**
Let x = Emily's brother's age
$(x - 5) + x = 21$
$2x = 26$
$x = 13$

Q4 **A**
Area of square = 15 cm × 15 cm = 225 cm²
Area of triangle = $\frac{1}{2}$ × 6 cm × 15 cm = 45 cm²
Area of Figure B = 225 cm² + 45 cm² = 270 cm²

Q5 **C**
$0.2 - 0.1 = 0.1$; $0.1 \div 10 = 0.01$
So each mark on the scale signifies an increase of 0.01
$0.1 + 0.01 + 0.01 + 0.01 = 0.13$

Q6 **B**
$49 \div 7 = 7$ so she needs 7 sets of 5 days to stitch 49 dresses; $7 \times 5 = 35$ days
$35 \div 7 = 5$ so she needs 5 weeks in total

Q7 **D**
Function machine multiplies by 7 then subtracts 2
Work backwards: $138 + 2 = 140$; $140 \div 7 = 20$

Q8 **C**
Number of coins = 176 g ÷ 4.4 g = 40

Q9 **E**
Figure D consists of 43 small cubes.
Volume of 1 small cube = 3 cm × 3 cm × 3 cm = 27 cm³
Volume of Figure D = 27 cm³ × 43 = 1161 cm³

Q10 **C**

Test 10

Q1 **B**

Q2 **D**
Number of girls that chose dark chocolate
= 48 − 3 − 7 − 8 − 12 − 17 = 1

Q3 **A**
Number of days in April = 30
Amount of water required = 30 × 3.5 litres = 105 litres

Q4 **D**
$3 \times 3 \times 3 = 27$; $4 \times 4 \times 4 = 64$; $6 \times 6 \times 6 = 216$;
$2 \times 2 \times 2 = 8$

Q5 **E**

7	2	3
0	4	8
5	6	1

Q6 **A**
Number of days in a week = 7
Number of days of the week that have 6 letters in their name = 3 (Monday, Friday, Sunday)
So the probability is $\frac{3}{7}$

Q7 **B**
ROSEVALE contains 4 vowels and 4 consonants so a ratio of 1:1

Q8 **A**
Range = 41.3 kg − 29.5 kg = 11.8 kg

Q9 **B**
Number of weeks = £52 ÷ £6.50 = 8 weeks

Q10 **B**
973 cm = 9.73 m; 9.73 m rounded to the nearest metre is 10 m

Test 11

Q1 **D**
5.07 m = 507 cm
507 cm ÷ 3 = 169 cm
169 cm ÷ 2 = 84.5 cm
84.5 cm = 845 mm

Q2 **E**
Matches not won = 3 + 6 + 5 + 7 = 21

Q3 **B**
Total cost of tickets = £24.50 × 25 = £612.50

Q4 **C**
Percentage of pages that do not contain illustrations
= 100% − 25% = 75%
75% of 520 = 390

Q5 **B**

Q6 **C**
The other estimates are either too small or too large.

Q7 **D**
$1 - \frac{2}{7} = \frac{5}{7}$
$\frac{5}{7}$ of the chocolate buttons = 15 chocolate buttons
$\frac{1}{7}$ = 3 buttons
$\frac{7}{7}$ = 21 buttons

Q8 **A**
Jean's father's age in 1 decade = $b + 10$
Jean's age in 1 decade = $\frac{b + 10}{3}$
Jean's age now = $\frac{b + 10}{3} - 10$

Q9 **C**
$1085 \div 7 = 155$
$155 + 10 = 165$

Q10 **C**
Total number of cars = 10 + 12 + 3 + 20 + 30 = 75
Fraction of cars with 2 passengers = $\frac{20}{75} = \frac{4}{15}$
Pie chart angle for 20 cars = $\frac{4}{15} \times 360° = 96°$

Test 12

Q1 **E**
7 out of 8 parts of the squash is water.
$\frac{7}{8}$ × 2500 ml = 2187.5 ml

Q2 **E**
60% of 180 cm = 108 cm

Q3 **B**
$\frac{4}{5}$ m = 80 cm
Area = 600 cm × 80 cm = 48000 cm²

Q4 **A**

Volume of cube = 50 cm × 50 cm × 50 cm

= 125000 cm³

Volume of 1 cuboid = 4 cm × 3 cm × 10 cm = 120 cm³

Number of cuboids that can be made = 125000 ÷ 120

= 1041 Remainder 80

So 1041 full complete cuboids can be made.

Q5 **D**

Angle on a straight line = 180°

Interior angle of a pentagon = 108°

Angle X = 180° − 108° = 72°

Q6 **B**

$12 - 5a = a - 12$

$12 + 12 = a + 5a$

$24 = 6a$

$a = 4$

Q7 **D**

Angles a and b are less than 90° so they are both acute.

Q8 **E**

10.35 rounded to the nearest tenth = 10.4

Q9 **A**

Basketball is the most popular and fencing is the least popular.

Range = most popular − least popular

Let the number of children who play basketball equal x

$22 = x - 11$

$x = 22 + 11 = 33$

Q10 **B**

He must take the 14:43 train from Brighton and so he will arrive at Haywards Heath at 15:06

Test 13

Q1 **E**

Value of each set of coins = 3 × 2p = 6p; 7 × 1p = 7p;

6p + 7p = 13p

£2.60 ÷ 13p = 20 so there are 20 sets of coins

Each set contains three 2 pence coins

Number of 2 pence coins = 20 × 3 = 60

Q2 **B**

Interior angle of a pentagon = 108°

Angle A = 180° − 108° = 72°

Angle B = 180° − 90° − 25° = 65°

Sum of angles A and B = 72° + 65° = 137°

Q3 **D**

Q4 **B**

50% more than 10p = 15p

Pay for 300 lids = 300 × 10p = £30

Pay for extra 100 lids = 100 × 15p = £15

Total pay on Wednesday = £30 + £15 = £45

Q5 **C**

Total weight = 625 + 135 + 25 + 275 + 30 + 50

= 1140 g

1140 g = 1.14 kg

Q6 **C**

Total number of children = mean × number of clubs

= 18 × 5 = 90

Number in Drama club = 90 − 24 − 11 − 30 − 13 = 12

Q7 **E**

Coordinates of A = (6 , 9)

x-coordinate does not change

y-coordinate is 3 units above the mirror line so reflected y-coordinate should be 3 units below the mirror line so coordinates are (6 , 3)

Q8 **C**

51.29 rounded to the nearest tenth is 51.3

Q9 **B**

$12n - 5 = 9n + 28$

$12n - 9n = 28 + 5$

$3n = 33$

$n = 11$

Q10 **B**

A clock can be divided into 12 sections, each representing the distance from one number to the next. Each section measures 30° (360° ÷ 12)

At 7.30 pm, the minute hand points to the 6 and the hour hand points halfway between 7 and 8. Therefore, the distance between the hands is $1\frac{1}{2}$ sections.

$1\frac{1}{2}$ × 30° = 45°

Test 14

Q1 **C**

23:50 is equivalent to 11.50 pm

Q2 **D**

Weight on left side = 2 kg + 50 g = 2050 g

Weight on right side = 3 kg + 500 g = 3500 g

Weight required on left side to balance scales

= 3500 g − 2050 g = 1450 g = 1.45 kg

Q3 **A**

347 − 119.7 = 227.3

227.3 + 456 = 683.3

Q4 **B**

A triangular prism has 5 faces.

Q5 **D**

9 + 3 = 12

Q6 **B**

(118.78 ÷ 10) × (118.78 ÷ 118.78) = 11.878 × 1

= 11.878

Q7 **C**

Cola drunk by girls = 6 litres

Total amount of cola = 9 + 6 = 15 litres

Fraction drunk by girls = $\frac{6}{15} = \frac{2}{5}$

Q8 **A**

A clock can be divided into 12 sections, each representing the distance from one number to the next. Each section measures 30° (360° ÷ 12)

At 1.30 pm, the minute hand points to 6 and the hour hand points halfway between 1 and 2. Therefore, the distance between the hands is $4\frac{1}{2}$ sections.

$4\frac{1}{2}$ × 30° = 135°

Q9 **C**

The battery is 20% charged after 3 hours.

Q10 **B**

Number of plants sold in ascending order: 34, 51, 98, 112, 160, 197

Median = (98 + 112) ÷ 2 = 105

Test 15

Q1 **D**

$243 \div 30 = 8$ Remainder 3

Q2 **D**

Q3 **A**

$15 - 14.88 = 0.12$

$15.12 - 15 = 0.12$

Q4 **C**

The question mark lies halfway between x and y

Value of the question mark $= (x + y) \div 2$

Q5 **B**

£8 = 800p = 80 × 10p so Alfred has 80 coins.

Total weight of coins = 80 × 6.5 g = 520 g = 0.52 kg

Q6 **D**

Coordinates of A = (5 , 8)

x-coordinate does not change

y-coordinate is 4 units above the mirror line so reflected y-coordinate should be 4 units below the mirror line; so coordinates are = (5 , 0)

Q7 **C**

The number formed from the last 3 digits will also be divisible by 8

Q8 **D**

Length of park = 12 m + 3 m + 3 m = 18 m

Width of park = 8 m + 3 m + 3 m = 14 m

Perimeter of park = 18 m + 18 m + 14 m + 14 m = 64 m

Q9 **A**

x^2 is less than 16 so the largest possible value of x is 3

Q10 **C**

Perpendicular lines connect at a 90° angle.

Test 16

Q1 **A**

1 foot is approximately 30 cm

1 foot is 12 inches

So 5 feet 4 inches is $5\frac{1}{3}$ feet

$5\frac{1}{3}$ × 30 cm = approximately 160 cm = 1.6 m

Q2 **B**

1, 4 and 9 are square numbers.

Q3 **E**

7 days in a week; 3 days have more than 7 letters (Wednesday, Thursday and Saturday) so the fraction is $\frac{3}{7}$

Q4 **A**

5.1 cm ÷ 3 cm = 1.7

Actual length = 1.7 × 7 m = 11.9 m

Q5 **A**

Q6 **D**

9 kilograms is approximately 20 pounds

So 81 kilograms will be 9 times greater

20 pounds × 9 = 180 pounds

Q7 **C**

Hours worked per week = 7 × 5 = 35

Money earned per week = 35 × £17 = £595

Money earned in 4 weeks = £595 × 4 = £2380

Q8 **B**

The first and second shapes have more than one line of symmetry.

Q9 **A**

Cost of 1 eraser in pence $= \frac{(3a - 1)}{12}$

Cost of 1 eraser in pounds $= \frac{(3a - 1)}{12} \div 100$

$= \frac{(3a - 1)}{12} \times \frac{1}{100} = \frac{(3a - 1)}{1200}$

Q10 **A**

Triangles, squares and hexagons can tessellate.

Test 17

Q1 **D**

Let x equal the number of red peppers

Number of green peppers = x + 11

$x + x + 11 = 45$

$2x = 34$

$x = 17$

So number of green peppers = 17 + 11 = 28

Q2 **B**

Scheduled arrival = 19:45

Estimated arrival = 20:08

19:45 → 20:08 is 23 minutes

Q3 **C**

19 × 120 g = 2280 g = 2.28 kg

Q4 **C**

Q5 **C**

Q6 **B**

$x - 6$ cm = 18 cm

$x = 24$ cm

Q7 **B**

Angle along a straight line = 180°

Each angle in equilateral triangle = 60°

Angle opposite Angle A in the parallelogram = 180° − 60° = 120°

Opposite angles in a parallelogram are equal.

So Angle A = 120°

Q8 **B**

Scores arranged in order: 35, 49, 96, 110, 110, 111, 123, 132, 167, 211, 333

Median score (middle of the ordered scores) = 111

Q9 **B**

Figure E rotates 90° clockwise to look like Figure F.

Q10 **C**

Sandra saved 20%

80% = £120 so 20% = £120 ÷ 4 = £30

Test 18

Q1 **C**

Total number of questions = 8 × 5 × 16 = 640 questions

Q2 **D**

Area = (18 m × 10 m) − (5 m × 2 m) = 180 m² − 10 m² = 170 m²

Q3 **B**

900 ÷ 48 = 18 Remainder 36 so 18 full cases are packed

Q4 **C**

Pentagon is formed from 5 identical triangles.

So area of 5 triangles = area of pentagon

Area of 1 triangle = $\frac{bd}{2}$ cm²

Area of 5 triangles = $\frac{5\,bd}{2}$ = 2.5 bd cm²

Q5 D

$(2^2)^3 = 4^3 = 4 \times 4 \times 4$

Q6 B

Each number is 0.55 greater than the number above it (and to the left of it).

So missing number = 4.9 + 0.55 = 5.45

Q7 D

8 × 13 = 104

11 × 13 = 143

So 8:11 is equivalent to 104:143

Q8 C

Angle X = 90° + 65° = 155°

Q9 C

Probability that the team loses = 1 − 0.62 − 0.3 = 0.08

Q10 A

Volume of 1 cube = 3 cm × 3 cm × 3 cm = 27 cm³

Number of cubes in Figure E = 20

Volume of Figure E = 27 cm³ × 20 = 540 cm³

Test 19

Q1 C

1649 ÷ 9 = 183 Remainder 2

Q2 A

A reflex angle is greater than 180° but less than 360°. The arrow shape has two internal reflex angles.

Q3 E

49:343 is equivalent to (49 ÷ 49):(343 ÷ 49) = 1:7

Q4 A

Range = 90 − 34 = 56

Q5 C

Q6 C

1 out of every 4 people has a trampoline = $\frac{1}{4}$

$\frac{1}{4}$ = 0.25 so ($\frac{1}{4}$ of 132) equals (0.25 × 132)

Q7 B

(1.2 × 5) + 1 = 7

Q8 C

The segment representing coffee takes up about a quarter of the pie chart.

Q9 C

Area of 1 triangle = $\frac{1}{2}$ × 4 cm × 6 cm = 12 cm²

Area of rectangle = 5 cm × 2 cm = 10 cm²

Area of Figure A = 12 cm² + 12 cm² + 10 cm² = 34 cm²

Q10 B

A = 101

B = 50% of 205 = 102.5

C = 50.75 × 2 = 101.5

D = $\frac{1}{3}$ of 312 = 104

E = 0.25 of 400 = 100

Test 20

Q1 B

Use estimate: 200 × 300 = 60000

Q2 B

Number of students = 3 + 8 + 22 + 18 = 51

Q3 C

Volume of cube = $2x \times 2x \times 2x = 8x^3$ cm³

Q4 B

Q5 D

Area of parallelogram = base × height

Area of triangle = $\frac{1}{2}$ × base × height

Q6 A

y − 7 = 15

y = 15 + 7

y = 22

Q7 D

j ÷ (5 × 4) = 1.5

j ÷ 20 = 1.5

j = 1.5 × 20

j = 30

Q8 D

Enlarged side length = 2 cm × 3 = 6 cm

Q9 B

The milkman delivers milk 6 days per week.

Total amount of milk = 1.5 litres × 80 × 6 = 720 litres

Q10 E

Number of metal objects = 29 + 30 = 59

Number of blue non-metal objects = 14

59 − 14 = 45

Test 21

Q1 B

25% did not complete the walk.

25% of 576 = 144

Q2 E

Each face of the large cube has 9 smaller cube faces. The larger cube has 6 faces.

Total number of smaller cube faces = 9 × 6 = 54

Number of smaller cube faces with a star drawn on them = $\frac{2}{3}$ × 54 = 36

Q3 C

49.58 rounded to the nearest whole number is 50

49.58 rounded to the nearest tenth is 49.6

Q4 B

The width of the lawn is 10 cm on the diagram.

So the length of the water feature is 10 cm on the diagram and its width is 7 cm.

Actual length = 10 cm × 120 = 1200 cm = 12 m

Actual width = 7 cm × 120 = 840 cm = 8.4 m

Actual area = 8.4 m × 12 m = 100.8 m²

Q5 C

624 ÷ 6 = 104

Q6 A

12 × ? = 1560

? = 1560 ÷ 12 = 130

Q7 D

Wool used per day = 50 cm × 12 = 600 cm = 6 m

Wool used per week = 6 m × 7 = 42 m

Q8 D

1 metre = 3.28 feet

29.52 ÷ 3.28 = 9 metres

Q9 A

Water left in carton = $1 - \frac{3}{5} = \frac{2}{5}$

$\frac{2}{5}$ × 2.5 litres = 1 litre

Q10 B

Annual Climbers = 48 − 13 − 11 − 10 = 14

Test 22

Q1   D

10 tree lights = 48 watts per day

45 rope lights = 3 × 120 watts = 360 watts per day

30 bulb lights = 300 watts ÷ 3 = 100 watts per day

Total power consumed in 2 days =

2 × (48 + 360 + 100) = 2 × 508 = 1016 watts

Q2   B

Each number in 2nd column is 5 less than the number above it, so missing numbers are 14 and 9.

Each number in the 3rd column is 6 less than the number above it, so missing numbers are 18 and 12.

Q3   D

9 m = 900 cm

After first cut, each piece is $\frac{900}{b}$ cm long

After second cut, each piece is $((\frac{900}{b}) \div 2)$ cm long

$\frac{900}{b} \div \frac{2}{1} = \frac{900}{b} \times \frac{1}{2} = \frac{900}{2b} = \frac{450}{b}$ cm

Q4   D

A square-based pyramid has 8 edges and 5 vertices.

So $v = e - 3$

Q5   C

Reflex angle between hands at 7:00 is $\frac{7}{12}$ of 360°

$\frac{7}{12} \times 360° = 210°$

Q6   A

Callum's share = £151.20 ÷ 3 = £50.40

Remaining share = £50.40 − £12.50 = £37.90

Q7   C

$6n - 9 = 165$

$6n = 174$

$n = 29$

Q8   E

Discounted rate = 100% − 30% − 5% = 65%

65% of £90 = £58.50

Q9   E

Q10   B

45.09 kg = 45 kg and 90 g

Notes

Notes